I0492283

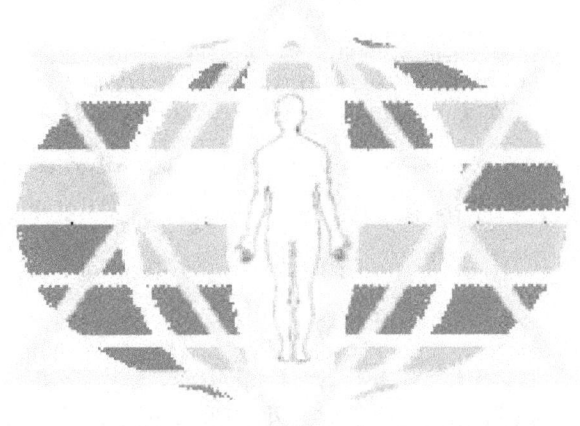

THE WAR OF DREAMS:
the battle continues

Dr. Obed Nuñez

DEDICATORY

To all whose avidity allows them to
seek beyond what the sight reaches
knowing that we are originating
from beyond dreams...

To all those whose inspiration
compels them to seek light in the
dual world...

To all whose believe in life beyond
life...

And to all those whose believe...
and elevates soul -> Spirit

INDEX

PRELUDE   1

COMMUNICATIONS OF SENSE   7

A LIGHT ON THE ROAD   16

OPENING DOORS   29

BEYOND THE DREAMS   37

WHAT IS NOT HIDDEN,   51

CONTINUES THERE

A WORLD OF SILENCE   62

TORMENT

AT THE END OF THE ROAD   81

EPILOGUE   93

ANNEX   101

PRELUDE

Some time ago...

Beyond communications. A world with communications organizes, feels and expresses itself ... it is built and inhabits every corner of the planet, without being able to... it feels interpreted... a world that appears in the collective imaginary with its own creation, which is created and constructed himself, hoping to be understood, within the reality lived and experienced by those who live in darkness, in the dark of night, to silence his moans... ¡But also, in the light!... They manifest, are there, orbiting between spans and spans, making itself feel more and more... those waiting to be discovered and reciprocated as inhabitants of the unknown, at lurk of the humanity.

1

How to do to express in silent? And say they are there; that live and coexist with one. With each one of this planet, in silent discouragement, is giving life to the earth, the electromagnetism... Who can it to understand? How many will have it to interpret? How to do to communicate the incommunicable?... What it's about is to describe the recent history of this great conspiracy, to make it present, to the light of humanity. Of those who need to learn, to know: What is happening to them? Why are you so tense? What excites them so much? What they are or are beyond them? And how does that encounter occur?

That communication of sense that exerts a significant influence on people and their subconscious world, but which is as real as the lived. This is the story that banishes, that fights for land... and all these

2

concerns are in the light of humanity for those who want to know the beyond, up close, and who need to know that what they feel is part of a concomitant reality that is expressed more and more every day.

Such a day as today, on a summer morning, I woke up with a lot of passion, with a lot of tenderness to describe, to decipher what I had dreamt. A world full of polychrome, full of lights, shadows, images and even sounds, which is created every time we sleep. And that when we are awakes it goes inadvertent. That world that we share every day with many others without knowing it. These are hours of life that are lost in that fade, in that meaninglessness of the sense of thought apparently ours, and of which we are always filled. Something that is part of life, of the mortal and of the common human being. A world full of lived fantasies that gets lost in the sunset of awakening, but that encloses much of its present existence. Its conflicts, its contradictions, its discouragements, its unloving, etc.

Word every time, full of surprises to discover, and that for some, sets guidelines in their life without knowing it. For others, it passes like a shooting star, flush, in the light of its path. What is there; vibrant, bold, and swift. Full of mess and entanglement, of duskiness, of obscurity. Or with grays and shades.

There, the symbols to discover of each one. And each one must do this personal work of discovery: his personal symbolism, dream-awake... It is as the poet says "walker there is no road, the road is made by walking, blow by blow, verse by verse".

As well, in the appeasement of the night, of the dream, there are many litanies, many ups and downs, and many horizons, full of lived passions that mark and govern the destiny of a person. That we are but dreams; dreams lived. We live in a present dream. Dreaming costs nothing, is part of life. It's life itself. So we walk on a path full of hope, crammed with discouragement, but we follow the path. We travel that road, as we walk through the world, full of passions: human passions. I dream to live and I live to dream. That's my world, the world that everyone inhabits. Since as we have understood, it is a dream, asleep or awake. It is an incessant dream, a wasteland, but I dream at last. Dreams that become a reality. It is so that it is entitled to "The war of dreams, the battle continues".

This transit, makes sense to all this: the world of the living, the one of the gone,

5

and that of those who are here ... Seen this way, the arrangement of the writing is conceived as: prelude; communications of sense; a light on the road; opening doors; beyond the dreams; what is not hidden continues there; a world of silence torment; at the end of the road; epilogue; and annex. And he does reference-mention concrete to SER UNO (Being One), the bible and to some sayings...

In this work, redundancy and style go hand in hand, in a coming and going... Between complexity and simplicity, with admiration, comparison and criticality... And it's not about dreams itself, but an approach to metaphor *dream-sleep-wake* in the dual world...

As the English teacher said, Keaton, in the movie "The society of dead poets ... only in his dreams can man be free".

COMMUNICATIONS OF SENSE

Today is one of those moments that present with much nonsense, full of ups and downs, and things without meaning, but with the sense of not knowing what is happening, or what is happening to me. To be trapped in a darkness that I would like to decipher. Voices of the most there, events and circumstances that appear at par, everywhere and that we do not understand. It is the path full of thorns and intrigues that every day become more lively and exciting, of jumps and scares. For the pleasure or panic of knowing and knowing what is happening to me and I am feeling. Something that perhaps is happening every day more, to a large and immense majority who do not know what to do or what to expect.

They are communications from beyond that seem more from here. And that we get confused to interpret them and presage them. And we believe that they are things that are making us crazy and that engender fear. ¿Is it something we do not know? But that it is there, and lives with me, and with you, and with everyone. Times are changing and every day they become more evident.

What is happening here and everywhere that drags us down an uncertain and dangerous path? What are you saying? ¡But dangerous of what!? We always engender fear for anything. And we give it a feeling and emotion that is not anything. And we associate it with a bad thing. And we think that everything is bad. Because, we are not ready for something and everything. To live and to die. And if is that we die? Or dying is being born? What happens to you then?...

This is the true story of what we feel in many lived moments of human life, to which we have no explanation, and stuns us, and startles us and unsettles us. How to fight such things? And how to decipher them? What is the beyond? And what's that? How do you sew that? Those seams that are broken and you have to go back to re-baste. What draws us here, to these omens and events of human nature that are more divine than human? Those impossible questioners.

What happens is that we are in communication with oneself and with others, in a different, diverse and contradictory way, and we do not know why. We have many doubts and questions that are not yet resolved; neither by science, nor by spiritual. Despite the fact that these times seem to be indicating something that goes beyond the material-spiritual human life that we do not know or are clear about. That story is there, day by day, more alive than ever. And we cling more to the material than to the spiritual. And we want to give to everything a material explanation. Being that perhaps we are not spiritual beings. Who inhabit and inhabit bodies, souls, and spirits. And there is the dilemma. The dichotomy it seems to be a whim, a calvary but it is not. It is reality becoming more and more present. More audacious, swift, to find us and get what we are. Souls in evolution in the human world, aspiring to elevation, through love. The love of all things and to all men, the people. It all comes down to a magic word. The immaterial-spiritual love.

And what is the concern that is about the famous word love? Love is what unites us, ties us in some way. That overstuffed feeling much malice, denial, incomprehension. But in the end, what we are is pure love, reflected in everything we do, we seek. What is, but to love truly true? And how much will he love, and how do we love ourselves? We are trapped with no way out, in the love. But how much we look for it, without ceasing? We have it there and we do not realize it. Because, we are more slope of other things than of ourselves and others. And of that fatuous and true selfishness that traps and corrodes the human being. That individual who does not look at the true real, what is hidden or what is present, but rather the vain, the fantasy, the impersonal-personal. As the poet said, "What we are... we are. A body of heroic hearts to fight and not give in"...

Now we live in a world of sympathy for the devil, for the gray, for the mean/petty/shabby, for the circumstantial and banal. Trapped in a world full of loneliness, and crammed with fantasy and bad feelings of the others for the others. That has been the history of humanity that has been done, every time, plus gradual and routine. But we must find the way. That's action true and forever... That essence of which we are made. What we are or are composed of: communications of meaning. However, sense with respect to what? And I say... To yourself, your guide, your superior self, to what moves us and elevates us. To this that is beyond ourselves. To what is truly... To what we really are, true... To the image and likeness of God. Almighty God who dwells in the heavens and on earth. That who made us and traps us in the sea of true happiness, the attainable and unattainable, but that is there and lives with all of us.

How do we find us then; to ourselves? In our true redemption, intermediation, and spatial-spiritual ascription. What are we

12

and what are we made of? That is the dilemma, because now they tell us that we are more than a physical body. We are a spiritual being, embodied in the material. That we are composed of soul and spirit. But until now, no one has been able to explain and understand certainly, as is that conjunction, really and truly. The hidden in each one, in each being. There are many artifices, gadgets, subterfuges, about it.

What it is about then; is to understand, those communications of meaning that are being given and expressing in a gradual and circumstantial way that reaches each one through different tracks and / or ways, perceived or not. Here is the dilemma again. We are dilettantes of the material-spiritual life, accompanied by many conflicts and contradictions. But the secret is, in the communications of sense; common sense to all flesh-bone mortals but not spirits. And many events are happening that are revealing the secret of the spiritual life.

The question is: how much you are or are ready? Because we're still green, in diapers. Only a few have been able to decipher it with the help of the others on the other side of the veil. And I think times don't give much, because we are just beginning, although some believe and promotion, the end of time, and the beginning of a new era.

There is a lot of cloth to cut, and a way to go. So the end in itself has neither arrived nor arrived so soon. That is just a problem of relativity of the things that many carries, and of those who live or want to live from the omens, intimidating others. Making of it, a way of living of the material that discerns from the true spiritual. Of the spiritual learners who still do not understand the differences of such action really and

14

truly, and who continue to make, interpretations of material office, to validate the spiritual. Trapped in this world of circumstances lived as a spiritual business.

But what is true, real and true, is that we are now at the beginning of the second decade of this century, more immersed in a spiritual world that embraces us and shelters us beforehand, and that encourages us, to a search for the real-true of being. That spiritual being that we all are. And with more encouragement and better techniques and instruments. Helped and interceded by the beyond. Something that even science will not be able to understand, but that does not date long, about its approach. Since if they are spiritual beings who do science, invested with a spiritual fragrance, then; how is it possible that we do not find each other, or they find themselves? It is a curious fact that is yet to be resolved.

A LIGHT ON THE ROAD

This episode, supposes: the relative semblance of a man who had a dream. Bring love to all men through welfare, so that their life would be more pleasant and enjoyable, in the face of so many glooms. Echoes of pain, bitterness, and no flavors in natural life, due to so many unresolved human-social conflicts that foster intrigue, envy, and greed (guts) for power. For dominating and subduing others, to their pleasure and egocentric vanity.

While we live hard and difficult times. Always attentive to circumstances. Oh! What a joy power to realize and get ahead. It is a constant struggle. It's a battle for life. But and what is life? But pain and suffering for the majority. Or is that: is there anyone who escapes such absurdity?

16

Because, whoever has the most, still; he suffers pain to maintain and preserve what he has. This is a world very materialist that has lost its way spiritual. And just a light begins on the road, back.

What happens to all of you is a shadow that lies in the spheres of the stolid mind that is confused in the bonds of pleasure without enjoyment. The accumulation of wealth, but of what nature, and does it consist of what?

We spend our lives - life - trying to accumulate material goods thinking that this makes life happier and more enjoyable. And it seems that we come into the world to suffer such punishment. And we don't realize how wrong we are. You are accumulating for what?... Who takes what he has? ... Nothing he owns and everything he learns and knows. That which lives and experiences, enjoys and produces lived satisfaction, achieved with effort, shared work, and compensated in harmony with others.

The material wrought, no! But spiritual yes! Knowledge, wisdom, loves for what he does and for others. The love of oneself, which is love for others. Love for everything. Shared and compassionate love. Love... a wise word but how complicated... because what is love? ...

The love is the source of life. I live to love, and to love to live. If as it is in the bible: God is love, and then, we are part of God, then; we are part of love. And we are made and composed of love ... to pure love, infinite love...

But in the meantime, many struggles and dalliances take place that tie us to

18

the mast of unfinished, capricious and meaningless life... And the rise and the sea, become a storm. And everything begins to disintegrate, to crumble, into crumbs of bread. Bread and wine, but they are not a path, unless it is shared. This is the story of every senseless walker who seeks and finds life, in that random whirlwind; instead of rising up with bread and wine, and giving thanks for what has been and lived. To live for life by living as God commands: with the love of one's neighbor. Oh! ... But what a difficult and elaborate thing to do...

Let's start by referring the anecdote of that man who succumbed to so much immorality and poverty. But not for that, it stopped being. I lived for the daily and the possible. Hungry and slandered by social life. That man implored time and misery. Save me, my God! He exclaimed. Looking where to grab on. And hopeful in changing their social status. The pain submerged him but he walked, like Ulysses, tied to the mast, not to succumb to the sirens' song.

Being poor seems to be a calamity and a social disgrace. As if that someone, it's the fault of being born poor. And nobody realizes that this is the social condition of the majorities in need, marginalized and oppressed by life and circumstances. The only thing that accompanies him is his faith and hope, which are the weapons with which the basic people are sewn, together with their regrets and hardships.

But all is not lost, why else? What is life for? For to live the moments. The simplest can become the happiest. And now, how do you find yourself, that man? Precisely in his dreams, in his most pleasant moments, in which he reaches crumbs of happiness. And how is that? I already said in the temple of Delphi. Observe yourself.

Of course, how profound for someone, whose social condition, already leaves much to be desired, after being an ordinary man who hardly accesses the confines of Greek philosophy. But what many do not know is that in every being, inhabits a whole laborious path and own philosophy, erected by the passage of infinite lives lived, in related spaces and times. This is hidden in everyone. And everyone discovers it in their own way. Because there is no magic or exact recipe. Since each man is a spiritual being, I have there, the secret of their seasonings. And everything has its time and its circumstances. Some more, others less, but each one gets their time and stays. Thus, each one is building his way and discovering his abode, where they deprive the goodness of spiritual life that adjust to material life.

So that this man, is stripping what is, and where it goes ... He is tilling with his free will, his way. And it comes out afloat, because it is discovering in others, its own pilgrimage. Which path he has to walk, to distinct and giant steps, while as he observes others, and he observes himself. And he struggles to conquer his inner self, in the reflection that is presented to him. Thus, he is learning and realizing. And already the man ceases to be what it was to be distinct and different. And it begins to connect with the social world differently. And it starts to look different. But in that hit, he also realizes that he starts to behave like those above and to criticize the foolish ones below... And he gets distracted and leaves his roots. And all that spiritual familiarity begins to fade because of the material. And he falls into the trap of those who so marginalized him, in his old social condition. Tremendous mess! Because it turns out like a skating. Now that you've learned, you make fun of those who fall... What forgetfulness! You forgot about yourself!

That is the story of many who go around 360 degrees but in the meantime, they seem to be headlong, face down; that is, invested in their original position. And the story of this man has been the novel of the known world, despite its progress and development. And it has had, which circle many times, in many times and associated spaces, to meet again himself, and back, pretending to... return to his Original position differently, spirally speaking, in another circumference. Look what he has had to roll. And hesitate to meet again, now, in societal conditions, different. The turns that the soul has had to give in the flesh. Everything is the oeuvre and grace of the embodied soul-spirit that dwells in you, but in another harmony.

The evil that dwells in you, in me, and in everyone, is a condition of the embodied soul, in evolution towards the spirit that dissipates slow and slowly; in a few more, in a few less. It is bad and good by nature. The more of one or the other, it is a serious thing. And it depends on the path that each one has traveled. And it is a relative thing in spiritual space and time, in its infinite hustle. Only on the way to love, it goes dissipating, such a comparison. In the nuances and the grays. Which side are you on, in how much and so much? ...

True those comparisons are odious, but it is the only way we have for now, to progress. And I'm not talking about material progress but spiritual progress. Or you did not know that everything is spiritual. And the whole problem becomes or derives from the material; of the vibratory frequency, and of the inter-degree-dimension zone in which you find yourself. Thus, each one comes to fulfill his spiritual life purpose. And you need a dimensional body, depending on where you are. And as the soul needs a body with which to express oneself, the events occur in the degree-flat dimensional that for now is third. But beware you do not go back, if it is that side and pole are talking, or forward, if that is what it is. Because everything is relative and circumstantial. If so, it exists. Because everything we want to see in the fantasy of the terrestrial human world.

*E*verything is confusing. Nothing is as it seems, or nothing is. It is only the reference: relative to itself; recurrent for the events that originate and depart from an event; and recursive, because it has the capacity to generate spirally, like an accordion, perhaps ... but one after another, according to the start of the previous one. And not time neither schedule neither space on the stage. It is from here and there and beyond. It is everything and nothing. Furthermore, it is the beginning and the end; it is alpha and omega. So everything is relative and circumstantial.

But also, it is the reflection of you, with yourself. We already said it, or not? It is not seeing to believe as we have been taught, but rather, believing to see. Because it's not outside, it's inside. It's up and down. It's on the side and on the other side. Furthermore, it's in front and behind. There is no definite or definite position, and neither is there a heaven nor hell. Neither truth nor lie. It is what it is, and only it is.

Everything else is way tale invented by humans. In a material world, of spiritual measurement. Because it is the souls in its path, who create everything on earth. Or perhaps, we are not all: soul-spirits incarnated or in the flesh, if it suits you better. Or is it perhaps, the earth is not the history of human evolution material-spiritual? And everything that is supposed to exist on unnatural earth has been invented by man. So what is man? A co-creator.

Thus, all beliefs are original to men, with all their valuations, historically speaking. And they have transcended from generation to generation. From evolution to evolution. From progress to progress. From measurement to measuring. From cause to cause. And only what we see most, are always, collateral and eventual effects. But true truths, there are not; only relativities. Although they are appearing and have begun to become evident. It's a matter of time, but a long time. It is not already, neither tomorrow, nor the day after tomorrow. Neither next year nor the next. It has not yet been said or specified. Nor will it be said. Or maybe if... But it's happening in the present tense. As the maxim says: "what is to be will be and what is to come will come".

OPENING DOORS

Everything seems to indicate that the doors are opening by leaps and bounds for some, and slow steps for others. What is certain is that something is happening, and happening in the light of all who perceive, how time is spreading faster. There are many omens about it. But what doors is it? That is the dilemma. And when? The bet is because it is soon. But for what? I keep asking myself. What does each person want to transfer, enter, and shorten? It is a difficult question for an answer, too difficult. Each one is outlining their way, at their own pace and pace. That depends on your free will.

What we are in tune with is in the form and the substance. And a lot has been written about that and a lot is raving too. Religions are plagued with it. The spiritual dogmas, doctrines, mystics, lazy and lazy (they live from the spiritual), from the occult, forecasters and prophesies; for say, prophets of disaster or rise, etc. The bible says: let the children come to me, for theirs is the kingdom of heaven! If we are all born as children, it is because we all come from heaven. And where is heaven? And hell? What is one or the other? Some believe that heaven is above and others believe that hell is below. Why this strange conjecture? If everything that goes up goes down and everything goes down goes up. Because, we insist on horizontalize or verticalize, on a flat in space-time. But it turns out that the one, who stands in the north, is inclined to stand. And the one that stands on the equator is horizontally standing. And the one, who stands in the south, stands vertically with his head down.

Because the balloon is round and oval, bulky and flattened.

There is not a fixed position but a relative one, since everything is in motion and dynamics. Everything rotates, turns on itself, and above all. So where is heaven and hell? Why do we believe that we go up (we elevate) and not down? Or it is the same relative position inter / within dimensional space. IT IS THE SAME PLACE seen where ... What is there, is there. It is multidimensional. You may not be discovering the warm water, because this has been said. I also do not try to discover anything. Everyone who discovers theirs. But it is handled confusing and intentionally by many to many. Above all, I rely on the pretext of faith and beliefs. As the poet said: "What we are, we are; a body of heroic hearts, with mind and destiny, willing to fight and not give in ". Trap size.

We have dealt with multiple layers of infinite life, ready to do anything to reach the alpha and omega, the beginning and the end. And in that pilgrimage we walk in a world of third dimension. We search, we search and we search. What to find? If we already have it and we do not realize it: Love. Where is the heart that I do not hear your throbbing, like the song? It is time and destiny. We return to the beginning. ¡What is love and what is it for?! Dilettante of dreams. So what are you looking for if you already have it? Shared dreams. But what are the dreams? In another flat? There is a false dilemma. The dreams, are they dreams? ... The loves, loves ¿are? ...

And you, what doors are you opening? Where is your heart? Reason deprives the heart, but there is no reason without a heart. We must unite reason and the heart.

Recently reunited in a very spiritual area (Sedona) of the most developed, dominant and oppressive country in the world today; a person, considered by many with an important degree of spiritual elevation, managed to preach that the tongue united to the palate causes the brain and heart to communicate in deep meditation ... Neither seat nor dissent from it, because it is the practice, in so much experience and living of each one who makes such an axiom, true truth or palatial frustration ... It is a material connection for a spiritual connection ... It is in the single, simple, where the greatness of the whole is, while the complex is what fixes the reality.

In any case, in the body is the spirit, the soul and the mind that are the three most sought after, read, idolized and passionate components of the human race. The mind is in the brain but also in the heart, in the

intestines, in the stomach, in the skin, in the cells, in the DNA, and in the whole body, because everything thinks and acts according to its condition to the unison of others, in occasional vibratory harmony due to the event or circumstance that occurred, in the evolution of life itself (material-spiritual). It is the soul and the spirit that give life to matter. In as much, everything has soul and spirit, in one or another vibrational degree. It is the spirit, the true light of life. And the soul, the means of transport that stores, keeps, conducts, permeates, by what enters and leaves, to the sound of the spirit. The spirit is the eternal and indubitable sound. It would be as if the orders spirit, soul leads, and mind-psyche program and manifested in the body-material present. And the body-material present through the mind, which feels and suffers, manifests to the soul that leads and asks permission to the organizing and computer spirit of

34

everything and the whole. Meanwhile, the spirit is in many dimensions, what it is: the essence of being.

So everything, lives and coexists in divine harmony, by the grace of the holy spirit (being elevated) which is ultimately, the own spirit embodied in presence and essence, in conjunction-union with the whole and for the whole, assisted by everyone. It is the evolved and elevated soul. Since elevation is a process of conscious conviction (awake), and it comes to soften the human creation and the spiritual-material co-creation for the survival and maintenance of the soul, and the empowerment of the spirit, while the alpha and the omega, come together to meet in their proper dimension spatial-sensory. At the end of everything, this is what we aspire to.

To say when it is the beginning and the end, it would be a heresy, because no one even knows ... And it will not be known by a few but by many, until then ... Which is also not the end of anything, but the beginning of another, that always come together ... It's a spiral, like geometry ... Perfect ... And that's the whole...

BEYOND THE DREAMS

There is forgiveness. This begins with the "Self" or the "ism", that is; oneself. Yet, what is forgiveness?! ... It is a serious thing. It is the detachment of something for something from someone. And that someone is your "me". Yourself. Which means, it is: get something from within condemns you and not let you live. And start with yourself. Only what is forgiven if you forgive yourself, then others, and others to you. It is a spiral of forgiveness. And it has many edges ... past that are present. It has to do with the cycle of life and past lives. It is an accordion in the coming and going ... from here and there and beyond ... it is the self many times, with many others, so many times ...

Forgiving requires the ability to forgive oneself, that is to say; recognize oneself first and then the others, and later... And to know apologize, supposes knowing how to love ... but love is something more... why! ¿How can you love another if you do not love yourself first? ... It is the "I" with me and then the "I" with you, here and now, and nothing else ... For I am "I" IN VARIOUS PLANS ... In the past, present and future, in the here and now, in The Life That I lived before (last), in now and Where I will live to return beyond ... We are talking about various dimensional plane-degrees ... That is the upward spiral of spiritual life.

This is Life. You have to live life intensely. Joyful is what rules life. Life itself. It is the life of one with the other, and the other with one. The pleasure of loving by loving what I lived. Mandino said, "Live every day as if it was the last of your existence." And the poet Whitman, spit: "Oh I... or Life! ... of the repeated questions... of the long train of perfidious... what is it serves of being among them..." And it responds this way: "That you're here... that life and identity exists ..." And add in the film Dead Poets Society: "That the powerful work continues... that you can contribute a verse... What would your verse be?"

Perhaps, because the life is like a verse that everyone composes in his own way, at his free will, at his own pace. So, we are all co-creators of what we think, feel, say, do, live ... And all in the same boat, in the same place, in the same footprint, in the same trap, in the same calamity, in the same heaven and / or on the same earth; is to say, in the same self with others and for the others ... you are not alone but in the spiral of life.

Do you live to die or do you die because you are alive? And then, what did I come for? That is the dilettante dilemma. Is it what you live once or many times? Would it make sense to live a single life? So, that we worry about good and evil. And what did not do ... Even more, what sense does it have for those who live little, or who do not overcome childhood, adolescence, youth, in contravention with adults; the sexagenarian, octogenarians, centenarians ...

The art of life is your work. Where are the teachings of Jesus of Nazareth (Formerly known as Joshua Emmanuel or Jesod)? Beloved one to another, as if you were yourself. Work well for yourself and for others, and do not look at whom. What follows is: be a server. That is the way. And then, why don't you comply? Why is, what is so much proclaimed not fulfilled?

This is Love. To which we have referred in previous sections but still subject to dubitation. Because: what is Love? Someone knows? It is something that many talks about and preaches about; they hoist, they ramble, they speak, they boast, but that in the history of mankind, has had little practice. And It is on and traffic has been folded war, hatred, greed, ego, distortion, disharmony, disunity, and stop counting ... And still is the most unfortunate way to behave and act, possibly; by the vast majority of human beings in this dual society, to the detriment of what represents, I have expresses, to feel and affirms the love. Although it is the center of the spiritual theme of all human attraction.

Love is a word that is categorized, perhaps, as the most important in human jargon, but by dispersion it is the furthest. I love you, you love me, we love each other, and I love myself. Because: How can I love another person if I do not love myself? It is the same as forgiveness. Forgiveness and love are two sides of the same coin, where the edge of a coin (its third side) is life. It is a trilogy: love, forgiveness, life. What is the first, second, third? I don't know. Put it as you want, while I do not know if there is order. ¡If I forgive you?! ¡I love! And ¡if I love?! Then forgive. ¡¿Oh no?! What do you choose? Because the life is a choice.

I chose to come, and I am here and now. That means that I chose to love and forgive. For love Christ died. And many have died ... And how many have died to forgive? The love is in everything: in what dwells, in all that is and is not. Like it or not, understand it or not. It is here, there and beyond ... It is in the sky, in the earth, in the water, in the fire, in the ether ... scattered in the universe but it is ... It is the infinite permanence...

In a biblical passage from Matthew, I have said that "God is love ...". So everything is God, insofar as it is love ... if there is no love, there is no God ... So everything; life is God, while forgiveness is God, and we are all God: love, life, forgiveness. And you, me and the others, we are essence, presence, power and fractal of God ... And that is the energy that you have here and now in this dimension...

In such a way that love is an expression that escapes the comprehension of the current human world, as it transforms all that is, in essence and consciousness and this is something that the human mind-psyche (soul) needs to configure. Love is constituted, generated and shaped in the active-creative spirit beyond this existential

43

dimension. Love is a heavenly polychrome, divine and sensory. So we still do not know how to love or of pure and unconditional love. In as much the love, is the true manifestation of the understanding in the knowledge of universal-spiritual elevation that exceptionally perfects every spirit-being in its concreteness, and which is properly referred to in the books of the ONE BEING (SER UNO). After all, love is the true intermediary of everything that is and is not. It is the true balance of everything and nothing. The perfect balance. Of the empty and the full one. Of the entity, object, thing, subject, unit, body, organism, particle, Nano, etc...

This is the energy. And again the question arises: But what is energy? It is light. And where the darkness remains. If everything moves, it is kinetic ... Then darkness exists in movement. What is compact is dark because it has condensed light, that is, the darkness is denser ... Therefore, the dark is harder, stronger in consistency ... and if God is in everything, then it is in the light as in the darkness ... The body that conserves life is darkness ... But take note: love is light and where there is love there is no darkness.

This is the search. But what search?! Search for what and why? Why do we insist so much on it? What did we come to look for and where? It is only on earth, in which we germinate. And what! Is it that there are no other places? Because here and not there, and beyond ... Many questions and a few answers ... And inconclusive, unconvincing answers ... There will be those who do and those who do not... Many derivations are argued in the name of God and religion or spirituality itself. It depends on free will and / or understanding. And of consciousness: of being here and now.

Any search and generate discouragement and frustration, but also perseverance. It is the need to dismantle something; to show something; and that something is something because to live and why to fight... It is as if life were a perennial struggle to achieve something. But reach what?! ... And what happens if someone does not have that internal-external conflict. Will beings exist with those conditions?! ... Again the questions float without answers a priori...

It is said of those who have attained nirvana: category expressed by BUDA (better known as Siddhārtha Gautama) that characterizes attains wisdom and perfect knowledge through the release: desires, and consequently, emotional control; control of excitation and tension; individual and collective conscience; and of reincarnation. A through meditation and enlightenment. That is, the "Complete spiritual liberation". And now, who or who achieve or have achieved such a purpose? ... And how, why and for what purpose? These are the questions we will have to ask ourselves. And the answers are in oneself. So find them. Because of whom one speaks of oneself. Of each one, according to their merit.

To meditate can be, by means of methods and techniques with a constant and persevering practice, inasmuch as it is supposes, to be in the center; in the equilibrium itself; in balance, in the self. That is, to meditate is to be conscious in the present tense. It is the meaning of an "Enlightened" state. And what gives you such a condition here on the present earth? By dualism, by the issue of

"Ego", "Egolatry" and "Idolatry", and "the superiority of one with another", and the act of "dominating one another"... Can someone: be "Illuminated" on earth? At the end of the day: is "enlightenment" the final purpose of reincarnation on earth? In as much as, the "enlightenment" can be understood as that process that characterizes the reaching of what was said before: "spiritual liberation". But, about what and why? It is very simple and very complex at the same time: *pure and unconditional love*. To God, to yourself, and to others. If "God is love", then what are we and what we look for? ... That is the answer ... The hard is the process ... Because is reached no so what if others do not reach ... At least, not in the contemporary land ... Not in this flat dimension...

*T*his *"The all"*. It is here and now. It's me, it's you, and it's them. There are the others: what they are not. And there are others: those who come and go... So they are all ... And if they are all, then this God ... And if so, then what is missing: pure-true and unconditional love. Because, ¿Who can say I have not loved, wanted, sighed, and cried for another? ... Not even the worst of mortals... because even so; being bad, you have enjoyed what you have done for good or bad for one or the other ... In this life, in another or another ... In this life, in another or another ... And the punishment, if it can be called that, will be to return-incarnate to supply his Karma-Dharma (weakness, difficulty, and overcoming through learning, joy and love) ... Within the universal spiral of pure and unconditional love, here, there and beyond ... And on the dimensional plane that is its own, according to its vibratory frequency-cyclical-energetic-corporal-material-spiritual... Y this is the concern of many or none ... Each one according to his

49

spirituality ... but that will reach them-he will touch, with all certainty this will be done...

WHAT IS NOT HIDDEN, CONTINUES THERE

You, me, and the others with their patterns-valuations-beliefs. Those engrams that entail commitment and difficulties at the same time, susceptible to be evaluated and overcome categorically: talking about images, sounds, colors, flavors, tenors, fears, tones, forms and backgrounds of circumstances and attenuating, etc. That is to say, of that one must review in his conscience-subconscious, in his inner self that causes him discomfort, disharmony and restlessness, of the now present and above all: of the before lived, in this life and in the others. Because, they enclose mechanisms that the mind and the mental spheres; the wrong memories, always bring you to collation. And in some disciplines, religions, rites, and groups, they usually have methods and techniques to solve-overcome it. And perhaps one of the most beautiful is posed by the "Ho Oponopono": *I'm sorry, forgives, I love you, thank you.*

You can get as much as you want from this Internet, or join or connect with a group or religion that considers and is appropriate and assiduous, according to your free will and your spiritual-personal search. But look for it and find it, because now it's about you and the others. And we are not alone, nor can we continue alone. Although keep in mind what we have left on the mat, in order to fail into other maneuvers: of manipulation and false prophets-masters-guides-charlatans-nickel. However, I warn you that these times are to get yourself, not to persecute prophets-teacher-guides, because you are your own temple-teacher-guide. Look inside, your answers that you will find there firmly ... Persevere, be constant, meditate, pray and act accordingly.

The path to wisdom is long, deep and complex, but perseverance and the constancy is a discipline to cultivate. And in as much, to walk in group or with a group, always it is desirable-good, and never is other, speaking collectively. Although his responsibility is individual-personal. Nobody saved anyone. Save yourself. ¿From whom?! From yourself. You are your own and strongest competitor-opponent.

The help you need is not of this world, as Jesus of Nazareth (The Christ) said: "My kingdom is not of this world." Search in your inner world for your inner "I". His God... And I would also say: "If you want to come after me ... take your cross and follow me ... I am the way, the truth and the life, and you do not come to the father but for me ... "But not exactly to him in person, but to the principles and legacy that I leave behind. And what does that mean? What does that phrase really express? The path is the light that is equivalent to love, and that is the true life, in your world, where corporeal materiality does not exist. But, the harmony, the balance, the subtle, continuity, correlation, vibration, frequency, rhythm, and more ... Where the fullness is the norm ... In degree, flat and adequate dimension, according to your soul-spirit meets with the ends of the categories-words, mentioned above ...

First Great Commandment: "you will love God above all things". First it is to love itself and then to others. But beware: do not be betrayed by your ego, and then believe that you are already enlightened, promoted and above others, and self-qualify or

54

pretend to be qualified as a teacher-guide, etc. We have already said it. While the second flourishes: "Love each other..." What a great and difficult thing, it has been for humanity, for me, for you, and for others ... See what is around you: in your family, your community, your neighborhood, your country, the country next door, from the north, from the south, from the east, from the west. Why are they killed? For belief-religions but also, for dominating one another. For appropriate their energy resources. As well by believing themselves owners of the truth, by *ambition*, etc. And for a great lack of those who allow themselves to be dominated-subjugated.

As it expresses the New Testament, in Matthew: "God is love and he who loves him ... loves him in spirit and in truth, what is of spirit, spirit is and what is of flesh, flesh is..." Then, the perfect equation: God =

Love = Spirit = True. Then, so what are they lacking? From love to the neighbor. That is to say, of himself.

Thus, what prevails is the meat, which is the material. But also, we must learn to differentiate the material from the spiritual. To the matter of energy. There is no matter without energy, nor energy without matter. When energy behaves as matter or vice versa: this is quantum. That is from God. Yours and mine, of the others. Those that are not visible but are. And they are part of the one and the other, and of the self, and of all.

A society-governed wants to solve their problems at the expense of others. This has been the history of the human-social world. And they did not have these predicates, two thousand years in the Christian era. So, how much time is needed to understand-understand it and act accordingly.

What has to happen? ... A bet on the destruction of the terrestrial world. Well, this has already been announced, and I'm not discovering or unveiling the lukewarm water. That is written in the apocalypse and in many books-scrolls and more, in the electronic books of BEING ONE (SER UNO), etc.

What are you waiting for, me, the others and the others who inhabit this world? Do not be a follower of anyone. Do not tie. Be a follower of principles and legacies that are worthwhile, strong and forceful, but put them into practice in true truth. Make it yours and others. Preach and practice. Be a militant of pure-true and unconditional love. That is love and God is love. Then, you are part of God ... God is in you, and you are a God in power, and it will be illuminated. And what is of God is not of men. We are talking about the new sup a man in another space-dimension.

In addition to this, we are answering that great men have come with great purposes-enlightenment that have arrived-fulfilled their role-protagonist on earth, in different times-areas-territories. And Western and Eastern history have recorded this. And there are, in these times convulsed of changes-transformations, and that are here, to help fulfill the purpose of the new times-era was in submission-mission. Some we have heard-known. And the fulfillment of prophecies, some found; other trapped; but prophecies in the end.

Nothing is deterministic, nor is it casual, because nothing has yet been said or determined with certainty. We must be careful with the prophets of disaster, because fear cannot be instilled in order to pretend that people become aware of what is coming, meanwhile, what is involved is defenestrating fear-afraid and above all: individual, and not less, collective. Since it is evident that we live in times convulsed by hatred, conspiracy and domination-subjugation of some over, in all social strata. And within society, the most patent sector of all is the politic; because he has always been the most perverse. And it can be seen, in an eventual world war in progress that the fourth is approaching, with unimaginable proportions, and that perhaps, it could happen sooner rather than later. And I'm not predicting anything. But it is broadly like the sword of Damocles.

It is public and notorious, what is convulsed in the Middle East. And it is precisely this region of the earth, one of the most important to carry out the exploitation of large oil reserves (oil is considered diseased energy, shameful and harmful) that ends up being, the fundamental energy flowing from Mother Earth (along with Venezuela). It is not an isolated fact, spiritually speaking that they are found in the bulging center of the earth, and as a spiritual connotation, the Judeo-Christian beliefs of the Western world, originally come from such zones, in conflict. Situation that has its greening also, in historical disagreements by cultural-religious beliefs. Misunderstood and poorly practiced, and distorted.

What we are witnessing then, is the end of one era and the beginning of another, as pointed out by the Mayan prophecies, and above all, the channeled-published about the ONE BEING (SER UNO). It is a new cycle that brings about a terrestrial rearrangement (of mother Earth), but also changes at the psycho-mental-spiritual level of those who inhabit this planet. And we have to get used to it, and achieve collectively and in unity that the transit is one of peace, love, harmony and happiness. So that depends on all of us. So each one fulfills its purpose, its mission; in greater love.

A WORLD OF SILENCE TORMENT

Today is yesterday as the future is today. There is nothing that does not turn or rotate in a spiral of life and sympathy ... everything flows from one place to another, if it is about sides ... What I am and is here is now, and now it is today-tomorrow ... Past has been in the today and now ... Everything is there and beyond ... Then, and of course, everything flows ... It continues its course ... It is a world of silence, torment. But what does this mean? It is to life and the world of the lived, but also to what we bring from other lived incarnations. What ties and cloaks us. Precisely what is here and now ... Let's to resurgence... We find ourselves constantly bustling from here and there.... Time is relative and belongs to this reality, and is measured linear, but not so, on another dimensional plane, with which we coexist-we exist in a soul-spiritual state. And there the tails come together in a circular spiral. That's why we speak of duality or dual world. What do I mean by this? That we are not alone ... That we exist in different plans, with each other and with the other, and with oneself, in its groping dimensional reality ... Known and unknown...

And there is a dynamic of action-interaction, representation with which we act-we intervene, in present-absent time ... It is a dimensional duality of multiple faces; With himself, and with others (and others), in the secret abode of the highest ... We intervene in dreams and not dreams, in the present moments of the here and now, and after the whole or in the whole ... In image and likeness of the creator, and with the universal creator ... One with the whole and the whole with the one ... This is the matter-spiritual thing. Because everything has matter and spirit in one form or another; the more of one of another that is relative to the dimensional space in which it is found- know-understand, locate-oblique...

*D*iversity of worlds, plurality of worlds... Jesus (Joshua) said: "My kingdom is not of this world" ... And it referred to the secret abode of the highest ... And why the torment? ... Good question it ... At last and in the end, the unknown begets this ... And the fear of the unknown as the death of the warrior, if we are all warriors in this land of God, which is yours, mine and others ... The same and himself, in the self, and with the same, here and there and beyond... It is fantasy, a dimensional allegory that alludes to the normal-paranormal and that flies- flutter in the air, in the skies that tell the story of human-non-human love and pain, an ascending spiral... Because that's where we come from and there we return...

Meanwhile, we see planets, stars, systems and constellations ... Infinite universe, creator and maker of all things ... Communications of sense-without meaning! For some yes, for others no. But that is the here and now ... The material-spiritual dimensional space of time ... As we walk and continue searching, listening and finding ... It is a matter of future terrestrial time ... And what we try to understand in the present tense: where did we come, and where are we going? And even worse or better: why are we here? ... Everyone seeks their own answer, if there is one. Because it is not yours, it belongs to everyone and for everyone, and it is on the way. We are coming soon but at your pace, time and destination, and we must know it, and find it in the future, because we will go and return, according to the cyclic wheel of the here and now, and of the for now...

Nobody has the concrete-present answer, because it is a process in construction-destruction, like all things human-social-spiritual: go, comings and to come... The passing (evolution, what is to come) of time... Or is it that perhaps you believed that I, and the others, or with the others, would have the answer... The only thing we aspire is to the message: to give it in the best possible way that is allowed, and perhaps to help it understand, as we try to understand it, ourselves...

This is a whole process of learning-practice-understanding-complement-arising ... that takes discipline, self-organization and respectful act ... As meditation-action and continuous work, with itself and with others. It is the act of love for you, for me and from me, for others and with others ... What have you been told, from the

66

time ... To your neighbor as yourself ... I have here the answers that you are looking for, and we seek... But it is nothing new! ... And then, what do you expect or better said: what do we expect ... Here is your (me) dilemma. Dilettantes of life, of the here and now...

*E*verything *flows*. Everything is movement. "Nothing is lost or nothing is destroyed ... everything is transformed". "The sun shines for everyone." It is the light of the soul that imposes itself and the spirit that rises towards its true origin and return. We are returning home. The somber and dark is diluted, and extinguished. And there is the bright crystal white light that we are made of (the spirit itself). What is it that we really are, what really remains in the existence of the infinite, what is the soul in spirit. In order to account, what is and always exists. The thinking thought-out of oneself (self). The thought-energy of which the One Being (SER UNO) speaks. It is the real and true infinite eternal life. The primary particle and cause of all things. What it is... But how is that?! ... There goes the process...

It is about the empowerment of the soul, when transitioning to energy-spirit. Remember: mind, soul and spirit, are the three great vertices of the triangle of lived existence. This is how the process is composed, and the three coexist, in an empowered way. What I am telling you is that matter (the dense) is redundant and has nothing to do with it. As a surplus, it is extinguished in the transformation, it does not disappear, but simply occupies its place in space, circumstantially determined by the spatial dimension to which it corresponds, its degree-plane. That is, it depends on the vibrational frequency to which the spiritual energy corresponds. If you do not understand, it does not matter, because I did not understand this, but there will be time to understand it, believe it and accept it. Know, feel, dissent and love, and in the end, feel and meditate, because only in its internal is the truth... Let's see, how I intuit-interpret-understand it.

The process of empowerment is incarnation - circular. Thus, from the earthly principle, the soul incarnates in human matter, many lives, to evolve, until reaching spiritual elevation in the dual world, which in this case is third-dimensional (earth = TERA). As pointed out by BEING ONE (SER UNO), thought-energies form the soul, and souls form the spirit. All this process is possible within the dual terrestrial world. In this bustle, the empowerment of the embodied soul occurs, in many earthly lives. How do I know? Everything has its time and place! It is your responsibility to seek it out and find it. Because it's not going to fall from the sky and nobody will give it to you, if you don't want to have it. The decision is yours.

It turns out that such individuality is merely spiritual fantasy, and more, of human-material condition. To say that we are individual beings, it is a purely social-organizational thing. On the material biological plane, many systems within the body universe, function as a whole-one: with very organs, functions and relationships, doing its energetic-material work, giving life to the incarnated body-soul so that the purpose of the incarnated human-soul mission to be executed may be fulfilled, in this dimension-plane terrestrial, according to what has been and is his existential plan, and in tune with his vibrational frequency, movement, rhythm, shape, color, sound, etc.

Assumed then, as has been said before that each engagement in order: subatomic, atom, molecule, cell, organ, system- organic, structure-corporeal, has its energy-thought quickens, gives strength and power, on each of these instances, generically speaking. As with the programming of DNA, it also responds to it. It means that for this to happen, everything thinks and acts according to its programming, at each level and corresponding priority, as an all-unit. And that united is what is possible, the individuality by indivisible, and that is where the meaning of the term "individuality" is understood.

However, like every concept it is that the bodily individuality is the simplification of the human-social understanding that it accepts that. However, as you can see, that is a potential universe: as it is above it is below, as it is inside it is outside. If one of the component-parts enters into conflict, programmatic confrontation-contradiction, it begins to generate chaos in the corresponding biological subsystem, generating the symptoms of the disease, from the energy-thought of the programmed part that enters into negative-sickness. That if not reviewed, rectified, and reprogrammed, and renewed, it becomes a chronic disease that degenerates into the corresponding bodily decay that in the end, causes the souls; to disengage from the body (material-corporeal death). And this degenerative process begins in the thought-energy associated with the subsystem-component compromised for various reasons, associated with a programmatic weakness-failure of the energy in question: in the soul and its evolution.

We refer to many energies-thoughts-souls, because that is what is involved in material-human life. Each person, when he or she incarnates his or her soul, brings all the experiences of their past lives, plus the thought-soul energies that are coupled with them from their parents, plus those that are coupled in the evolution of their life from the earth-humanity environment. Conforming to whole in its action, by vibrational similarity. It is what gives life to that bodily being. Meanwhile, what we approach with all this is that we are not alone, and we are not only as we believe, but we are one in the unity of the whole, bodily materialized. This is how we live and live together. And so we think many in one. That is to say, we are influenced by the thought-energies that swarm in the terrestrial sphere of electromagnetism. When we breathe, it is the same air for everyone, so it happens with positive or

73

negative-sick thought-energies, according to the soul filter of each person.

Therefore, we are a body-walking universe of organic-biological systems energetically incorporated, thinking in their parts-components, through a mind-one that regulates, directs and controls energies-thinking of souls chained in a subconscious-conscious, in a body to do-experience-progress-evolve in your current life. And that will be influenced by the thought-soul-energies that you have energetically speaking; feeling and living, an experience of life, to make room for the evolution of the soul.

In the end, what we want to say; is that we are one, meanwhile; a collective-group of energy-thought-soul in evolving, rather than an individual as such (at the time of what Carl Jung said about the collective unconscious that is the common area of experimentation to which all humans access beyond consciousness, independent of time and place, through the psyche, and that surpasses reason). Although apparently expressed in this way, in the material dual world, because it has already been said that we are mind, soul and spirit.

A soul is in evolution, until the spiritual elevation is reached, which in the end, turns out to be the spirit or elevated energetic being. And this is the process to which we referred earlier, that cyclically spiral, we are building in each incarnated life that we travel on this planet earth. In such a way that the goal is to reach the level of spirit (Holy Spirit of the Christian religion, nirvana, enlightenment, among others) that for now is still evolution of the soul. This is supposed to live many lifetimes to evolve and forward, until to elevate to the degree of spirit, which we still do not have. And that is the challenge.

I want to make it clear that if we do not reach such a level of spirit, we will continue disembodied-incarnating. And what emerges in the state of bodily death, it is like a bunch of grapes that uncouples from the bunch, and they go out directed towards a dimensional space. To give these energy-thoughts-souls will go: some to a process review-correction-reprogramming-renovation and incarnation of the disembodied soul itself. Other of those that made up the energy-matter-body (inserted in the subconscious-conscious mind), they are reintegrated, in some already incarnated-established soul. And if the soul had reached its first level of spirit, then, it would no longer need to incarnate more, having achieved its existential purpose in the dual world (according to BE ONE or SER UNO, available at www.elseruno.com).

Since part of the foregoing, subject to my knowledge-understanding, it is in tune with what inspired-channeled about Being One (SER UNO), and invite them to contrast-validate-understand-confirmed through books electronically supplied by its official website. What touches for now, is to make me answer with this genius-generality, as intermediary-messenger. And it has been done.

What I want to make clear in the end is: a mistake to believe that soul and spirit is the same thing, unless the evolved soul has reached in many embodied lives in this dual world, its level of spirit. In which case, amount to another plane and not necessarily embody more. On the other hand, it is a process that runs until reaching, such a level. Beyond that here.

The spirit ends up being a higher level of the conscious soul. It is the transit from the subconscious to the conscious. And it is in the subconscious where are the emotions that according to the One Being (SER UNO) are what we must make conscious, through a self-reflective analytical process, psychologically speaking. Work that is proper to each person in his free will and in his need. Because saviors there are not. And nobody is going to save them, while there are no messiahs or teachers who can or want to do their spiritual work. Only you are saved too. You are the owner of its causes and effects. When you see yourself in the mirror, the one you see there, is the cause of their actions and destiny, of their successes and mistakes, of their virtues and defects, and is responsible for everything

78

that happens to them for good or for bad. That is your god or your executioner, because that is you. And if you want to blame someone, then blame yourself. His is what Hindus call karma-dharma.

When you feel something against another, then check if what you see there is part of your mirror and by reflection, it is also yours. Make it aware, correct it and be ready to let it go. That means that I extracted it from the subconscious and by making it conscious, I release it because I forgive it. Otherwise, it is still there, it will be repeated or returned ... But do not worry, but take care to "realize", and carry out the process of self-analysis-reflective-critical as well, of the cause that originates that, so that the effect happens, by making it conscious.

Follow the clue to the thought that generated it. Unchain it recursively, and you will see that it function, while it becomes alert and awake. That is a good practice to unload the subconscious, of many traps, tangles, wrongs, injustices, "sausage roll", rolls and others that they already have hidden in the dark recesses of that labyrinth of their being. So that you can be your own psychologist, before these times of stolid minds, dissociated and charged with energies-negative-thoughts-ill.

AT THE END OF THE ROAD

Close encounters. You live in your material-spiritual world of different types. Some more credible. Others more passengers of strange coincidence or "DÉJÀ VU", and most without realizing it. You live to live and you love to love. Love without being loved. It is first, to love oneself, and then to others. We have already insisted on it. Loves who can, not who wants. Others will learn to love you after they love each other. That's it. And everything comes together. It is the union and communion of the whole for the whole. In its splendor, at its rhythm and color. Such are the encounters with oneself and with others, full of course: of their own ravings and contradictions. But what must be basted and categorized in the understanding of the encounter with life and lived. It is a reflection with itself, perennial and present, and with others, in the present body. Not more than a perpetual learning while you are here, and without a doubt, when you are in the afterlife.

After all, it is a process of learning the subconscious to become conscious in body-soul present. It's finding yourself. It is the light that becomes flesh and wine. It is the happening of the here and now. It is the hesitant and meditative, in inner-outer conscious reflection. It is the maxim of being here and now. It is the encounter with your light that is your individual salvation, achieved by personal merit. It is that nirvana that is so much sought after and yearned for. He is the dilettante of life for one's life and with the others who are us. In a spiritual-material conjugation of forces of action-reaction, in search of the true truth, in the unlucky walk of encounter-disagreement.

Above all, be aware that the battle continues, while you are here and do not let you wrap up for her. It is learning to be a militant of the conscious observation of the daily occurrence and becoming, so as not to fall into the networks of darkness-evil that subjugate those who live the dream of reverie. It is ultimately the awakening of consciousness. And make yourself always present, from the living soul-spirit with its own light, day by day, moment by moment, until the end of the path that you have to travel... And you'll see if you come back ...

Thus and all, it is your light and spiritual salvation, in the corporal materiality. It is to make way to walk and see the path that has to be drawn, as the poet said. Because it is a poetry that builds itself, in the dynamics of the metaphor of life, with full knowledge of cause and effect: reflective, recurrent, referential and recursive, in the present body. And that being present, I understand and comprehend in its actions, and therefore, I am able to make it known to others, starting with mine and those close ... I clarify: the reflective is the self (with myself), recurrent the consequent, disciplined and orderly, loyal and committed; and that is what refers to everything and with everything (be relationship cause-effect-mine-people-things, in time-untimely, etc.), while recursively comes and goes (as a chain), from the beginning to the end and vice versa, and intermediate, in thinking-acting-walking.

That is, full dynamics and movement, because everything flows ... It has already been said: nothing is static, because it does not exist in the universe. As there is no straight line (mere representation-relative observation), since everything is curved-circular ... It is the sacred geometry of the universe, of life, etc...

When you are here, is it for what? ... To suffer-endure? This is the maxim of many religions: the blow-stumbles teach and people have become accustomed to it. Blows, blows and more hits ... Stones in the road ... Errors and more mistakes ... Misunderstandings and more mistakes ... The survival of the fittest, the strongest ... The domination of a few over others, in majority: subjugation. All this is potential imperfection; to be restrained by hatred, selfishness, ambition, pleasure and more ... This only exists at this level the material-social, in this existential plane. This is what feeds the reptilian-energy which is sick negative thoughts supplied-fed by reptilian beings-entity. Because their subsistence depends on it.

A cold-blooded animal is called a reptile that, due to its appearance, has a rigid, hard, cobbled, scaly body, a prominent-protruding shell that crawls and satisfies its animal instinct, predator, regardless of what / whom, susceptible to appear ugly, hostile, and merciless. And there are people who show themselves-manifest-act and express in that way. In addition to that there is a whole theory about the reptilian brain, associated to similar behavior between humans and animals, and the instinctive thinking of surviving, in as much, certain emotions.

The closest reference to the characteristics of reptilians for their wrongdoing is relative to example to: political-parties-governments, corporations-companies-business, religions-dogmas sects, criminal- malfeasant-thieves or others who are similar in these world terrestrial social predators. Evil feeds evil. And it is something that is not seen, but feels and suffers. And there are more of those who cohabit with this than with its opposite: the good. Even though many, falsely preach, a supposed good, and thus abduct unsuspecting-worshipers or others. And all this is considered, the darkness in its manipulator-subjugating process.

What we wanted to sit with reptile/reptilian? ... That all without exception, we have a reptilian/reptilian part living inside that binds us and undermines us. The more or less you will have, it depends on your vibrational frequency and what your subconscious holds. You decide if you feed your subconscious, accumulating more energy-thought-negative-sick, or conversely, is releasing, rinsing, cleaning, or reprogramming them, making them ever more aware. To do this, you must do the work-process of becoming aware, as has been indicated in this manuscript, with the modes referred to, during all these passages...

It is essential for me to notice that there are beings that live-they coexist with this appearance in the universe, by the development of thought, more on the left side of the brain, and by lack of thinking on the right side, but not necessarily, are considered bad or predators per se, and they are known as reptilian beings. But in addition, it is said that the beings coexist-cohabit in a dimensional-plane on earth, whose evolution is held back-stacked, and these beings; subjugate humans by the left brain so that they lend themselves to their malice, feeding the negative-thinking-diseased-energies. And they are the axis of evil on earth (I suggest you investigate more since it is not the subject of this writing).

In any case, you are the owner of your faculties, and you decide if you delegate them or develop them. It is your decisions-works that will incite good or evil. Remember the saying that says: "to whom a good tree comes, a good shadow shelters him." And in as much, the biblical verse on which "by its fruits you will know them" (Matthew 12:33). And besides, works are loves, if they are good... Thus, everything ends in profit or loss, by correspondence and / or merit...

So much has been spoken of the destiny that has been a maxim that deprives you in the life of the people in the future. But you build it yourself, because it would not make sense to be predestined. There may be preconceptions that you bring from your experiences inherited from many incarnations, but in principle, you are here to overcome them-to improve them. Since we talk about change and the dynamics of the whole, nothing is said or nothing is unchangeable or indelible. While we bet on the evolution and elevation of the soul-spirit. And that's what we come to, not something else. No backtracking if it makes sense to stagnate. Of course that will depend on how you face the causes and effects to which you are exposed, in your daily-material-spiritual life.

In the understanding that you will be at the beginning of the incarnate life, subject to the mother-father legacy-life and the framework surrounding-environment-living. What you bring is yours and that is what your initial conditions will be, your supplies. But the process is your mere responsibility-creation, under the relative conditions of the bio-psycho-social-spiritual existence that surrounds you. And all these considerations-interactions are subject to the access-connection-place of the higher and/or planetary consciousness.

This is not easy to digest, because it is a work of knowledge- understanding material-spiritual, as theory-practice-experience, "realization of" through of the consciousness. That each person in his space-time-moment will be getting-understanding.

Each one will have its lifetime to live their spiritual characterization consciously or not, but not one escapes it. From each person according to his capacity, from each one according to his circumstances to know-understand what precedes him-happens.

You will also have, new lives-incarnations to complete the unfinished, in the temporary way of their unresolved   energies-thoughts-emotions, either by itself, in their next incarnation and/or through other energies-thoughts embodied in the corresponding side-zone (according to the Being One: SER UNO). And here I prefer to invite you to inquire into the electronic publications of BEING ONE, rather than recklessly making interpretations of what is postulated there. Although I maintain that I am in full harmony with what is deepened by them, based on what the messenger-contact-channeler,   have provided us through the Ayaplians brothers.

Finally, "The battle is with whom?" Answer: with oneself. You are your own contender.

EPILOGUE

It has already been said that life is a process of cause and effect. Everything originates from something already happened. It is the awakening of consciousness that counts and that is the challenge of being here and now. We see them every day of life, while it lasts, in that dilemma. Like it or not, if you agree or disagree, if you understand it or not. But that is what we have come to, and that is why you are here and so on you will come, until the end of time... And it will have to be fulfilled at some time universe-terrestrial. And this has already been proposed and estimated. And it is happening. From each one depends on reaching sooner or later, that present moment that is neither time nor distance, but perennial.

Life is like that, a game of present destiny. A going and a coming within the universe, because it is infinite. And it is not density-matter, it is simply energy-thought, always alive and existing, with or without a present body, if that is what it is about. It is the truth in essence and constant presence. It is what exists. It is what it is and nothing more. It is a permanent flow. It is a spiral of ascending energy and raising the vibrational frequency, to a rhythm, color, shape, correlation and continuity, and more... For something Descartes I leave sitting that "I think then I exist". I was right. As I always think, I always exist... And that's the omnipresent permanent: *the thought-energy*. And that is the flow of God, his God, and my God ... The God of all, of them, of the others ... That is the living God... The unique principle, the supreme cause of all things. So it has been said, so be it, and so it will be.

The War of Dreams, as this writing has been called, is the expression of thought in variable time-particular terrestrial. It is a maxim of material-human life. It is the unity of opposites: action-reaction, good-evil, light-darkness, constant-change, static-motion, continuity-rupture, negative-positive, the whole and nothingness. Ebb and flow. Soul and spirit. It is the dilemma of life, for which we are trapped here, since ancient times... And that is why we want and must leave. Back to the origin: where we come from, where we are from. What we are actually made of.

It is the awakening of consciousness in its true universal-stellar reality. The wake of light. It is learning-understanding what we are and what we are truly made of. And how to return never to return ever. It is to fulfill the purpose of non-incarnated life and to continue in the awakened universe.

In this earthly bustle, it is very easy to go back to sleep awake. And that's why we fall in the wobble of darkness. And we stumble on the same stone. And we fall. And we sleepwalked on pedestrian paths... In the trap of subjugation-manipulation, we are repeat offenders, and we reincarnate. We enter the same wheel of pleasure and destiny. Again the vicious circle. What are we doing wrong? The same litany. Why is it hard, staying awake?

Consistency is one of the maxims of spirituality. The latter is "a way of being, living and existing", as expressed and channeled by Franca of the older Ayaplians Brothers (Being One). Said thus, perseverance is an important and preponderant factor in all human activity, and not less, within the scope of spirituality. Continuity depends on much, reaching success. And the success in the spiritual path is to be constant and consistent with oneself, with honesty and sincerity.

Of course, within the limits of human-material reality that surrounds us and tramples, to take the out of the true spiritual search that is within each one. And it makes us doubt and pigeonhole or fantasizes about what is outside. Sometimes by tiredness, by annoyance and weakness, or by not finding, what has not been given to us. Because neither we have yet won it. Or for aspiring without meritoriousness. And we desist. And we become followers again... And we fall into the dream of reverie.

So the Christ consciousness must be kept awake. The spiritual principles intact and persevering. The awareness of

BEING, of being here and now. To be consistent with oneself and accept that everything arrives by deserving in time and destination. The spiritual practice must be a variable-singular in space-time, within the human-social reality that you have to live-experience. It is a matter of will, force, power, knowledge, understanding and energy of the soul. It is the true search and transverse-intertwined transit in love. This is the tissue of love. Love yourself and your neighbor. That is the light of each one and the one that illuminates the world in turn. Devoid of ego-selfishness and more...

When we have reached this limit of the message, then, we know that we are not saviors, nor magicians, nor teachers, nor channelers, among others of what we usually qualify, but; simple messengers of good, of positivity and love that like any living mortal, try to get everything that has been postulated, in this commitment to reflect on voice high.

Thinking in discovered-thought-described and within the life purpose of human brothers and sisters, on the path to spirituality. From my own personal transit, and for anyone who finds this useful, in order to shed light where it still persists, a certain degree of ambiguity-darkness.

Knowing that as a militant of spirituality, this is a way to reinforce my own transcendental disjunctive, in the transit of the path traveled, in this human-material life that still suspects the fantasy-like appearance of all that has been said-written by/about the human-animal history, by those concerned- obfuscated-manipulators-heartless in confusing-distorting, the true contents of the ventilated moral-spiritual principles.

The only way to confront, already entered the twenty-first century, with all the changes that have taken place, is its own true truth-interior, where is the ancestral-present legacy that will direct its personal-spiritual path. Decidedly, we are militants of spirituality and dilettantes on the road to I walk...

Knowing how to discern objectively is in the *immunity of the soul*. Everything becomes of my interior. And so is my belief-creed-create... Betting I understanding, I have been redundant in variety textual, and somewhat unruly in the background / documentary style.

ANNEX

This is what is considered the MERKABA, the six-pointed star, the star of David, two inverted pyramids, two inverted cones, torus or torus tube... that contemplates the six elements of the ascension (in each angle) that will have to be reached in order to be able to rise and come out of the third dimension, towards the corresponding higher plane and dimension. According to how each person vibrates, and from the bubble sphere origin of which it became, but also, relative to the vibratory frequency, in which it develops/coupling.

Thus, in the bustle of your earthly lives and until you incarnate anymore, in your passage through the electromagnetism of the planet, and as it is arranged in the Inner City of Crystal White Light, it will be where your ascension-programming ends and is reprogrammed, together with his superior being, and in harmony with the Elder Brothers of Light and Love, of whom he is a part and / or is original.

This supposes that the four elements, reaches them in their terrestrial lives, but the Ether and Helium, acquires it in the Inner City, to evolve and elevate. And this is possible with that body of light that we are trying to graph, below. What makes room for the Soul to develop its Spirit. And that is the only way to leave the planet and not another, according to BEING ONE (SER UNO).

In a certain way, what also reflects, the figure below, is the passage from the age of Pisces to Aquarius, that is to say: transit from water to air. It is the process that begins and is being lived, in this second decade of the twenty-first century (already prophesied, in times past).

Therefore, the movement of the curved arrows for the water-fire-ether triangle (seen from the back, water-shoulder-left towards fire-shoulder-right, by the impulse of ether), would be equivalent to a volume with exit point forward, relative to evolution, that is, odd-earth.

For the triangle-cone-inverted air-earth-helium, the curved arrow shows the flow with the exit point backwards (hip-right-air towards hip-left-earth-> ascension) that includes the elevation and final exit of the planet, that is, ground-even. (See image on the back cover).

Also, the figure can be of analytical utility to review the reflection of its own personality-action-correction, when fighting

103

battles with itself, on its subconscious-action-reprogramming, and reference with others.

From where you perceive yourself to clarify matters of your own human-material life, in tune with the spiritual. So, you need to know: in which element (s) of the four main is located, either by your zodiac or another.

Since fire and water are contrary, as well as air and earth. Meanwhile, fire and air are mirrors, as well as water and earth. And fire with earth, they are causes and effects, and vice versa. And in that same tonic, air and water, they are also causes and effects, in both senses. In this way, successively, you can make relationships and draw conclusions, about your life circumstances and what things work and / or improve, through correlation and correspondence...

It is your duty to contrast and validate what was said before, in different ways, to understand it in depth. Since what I am showing, is only, by way of information-intuition-interpretation, and not in any way, of affirmation. Because it would contradict, what has been exposed in the message of this work, and that is not our intention.

Star six-pointed or David, MERKABA, when it becomes 3D (next Page)

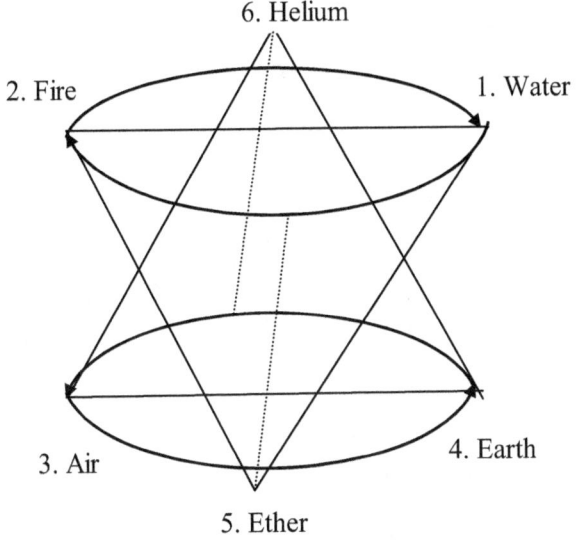

To Being One information, with elaboration and interpretation itself

www.ingramcontent.com/pod-product-compliance
Lightning Source LLC
Chambersburg PA
CBHW070608220526
45467CB00003B/1340